GPO CREATING A SUSTAINABLE FUTURE

I0040936

TABLE OF CONTENTS

ACKNOWLEDGEMENT

I would like to give reverence to the spirit of nature and our ancestors for providing solutions in a time of need. Also all of the good people who see the need and benefits of the eco-based changes such as Growing Pharms and Gamphace Research team.

INTRODUCTION

GPO (Genetically Purified Organisms) is a Growing Pharms made company that utilizes ancient growing processes passed down from tribal ancestors of the Mound builders and susquehannock to build and maintain a sustainable future. In order to specialize in the natural development of all plants, industrial ones like hemp and cotton are also those used for human consumption. Such as cannabis in conjunction with the GP CSA 885 a1 exemption for Research and development. This company will implement education of the use of natural resources to increase carbon sequestration, mineral levels in plants and food in order to invoke a healthier mind, body and environment. These steps will reduce carbon in the atmosphere and will dramatically decrease the effects of climate change. While also changing and reversing the negative effects of other plants and food that have been altered. Which studies show negative effects to the body, mind and world over time.

CLIMATE CHANGE

What is Climate change and why should we pay it any attention?

Climate change is the global or regional climate patterns, a change apparent from the mid to late 20th century onwards and attributed largely to the increased levels of atmospheric carbon dioxide produced using fossil fuels. A fossil fuel is a fuel formed by natural processes, such as anaerobic decomposition of buried dead organisms, containing energy originating in ancient photosynthesis. Such organisms and their resulting fossil fuels typically have an age of millions of years, and sometimes more than 650 million years.

There are three types of fossil fuels which can all be used for energy provision: coal, oil and natural gas. Coal is a solid fossil fuel formed over millions of years by decay of land vegetation. To put it in Layman's terms There are both natural and human sources of carbon dioxide emissions. Natural sources include decomposing waste and even bodies, ocean release and respiration. So even something that we need to live can negatively impact the plant. As well as the burning of fossil fuels like coal, oil and natural gas. That means the cars we use to get to work, the factories that make the cars and food amongst more. So, in a world that's constantly changing and evolving, how do we create a sustainable environment for the future.

As an American Indian Moor our tribes we are taught
to respect our mothers and fathers so our days may be
long. Most people don't realize that goes beyond your
physical parents and even equates to nature, Meaning
mother Earth. We have a responsibility as the inhabitants
of this plant to ensure it can sustain us and future
generation while keeping balance in the ecosystem

The United Nations concluded there's a more than 95 percent

probability that human activities over the past 50 years have warmed our planet. The industrial activities that our modern civilization depends upon have raised atmospheric carbon dioxide levels from 280 parts per million to 412 parts per million in the last 150 years. 95 percent probability that human-produced greenhouse gases such as carbon dioxide, methane and nitrous oxide have caused much of the observed increase in Earth's temperatures over the past 50 years.

Climate change is one of the most pressing challenges facing humanity, and agriculture feels its effects in profound ways. Farmers are particularly impacted by extreme weather conditions, which include drought, severe heat, flooding, and other shifting climatic trends. These all pose challenges for farmers as they work to grow enough food, which is why we're devoted to finding ways to transform agriculture to be part of the solution in addressing climate change. Although agriculture is a contributor to climate change, the industry plays a role in curbing greenhouse gas (GHG) emissions like carbon dioxide, methane, and nitrogen oxide that contribute to climate change

Simple Solutions

Carbon Sequestration
Is the key to a sustainable future and the proper development for technology agriculture, architecture and a balanced ecosystem and life!

Carbon sequestration describes long-term storage of **carbon** dioxide or other forms of **carbon** to either mitigate or defer global warming and avoid dangerous climate change. It has been proposed to slow the atmospheric and marine accumulation of greenhouse gases, which are released by burning fossil fuels.
So, the key is to reduce the use of things that emit co^2 and capture the existing overage of Co^2 in the atmosphere and seas without giving up our creative edge and limiting our technological growth. Sounds near Impossible right. Wrong and my ancestors knew this and applied the information.

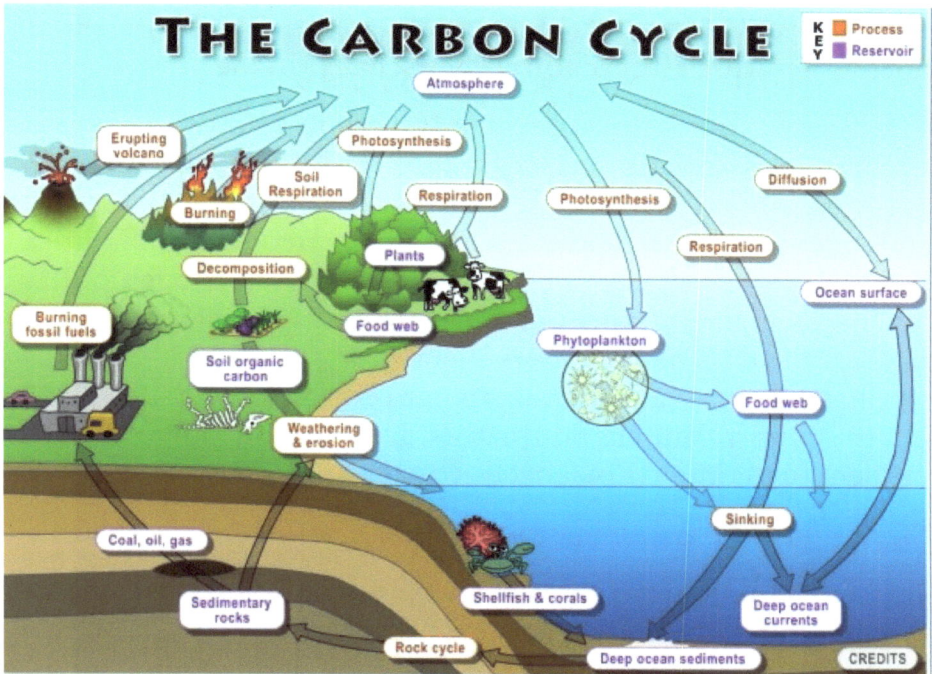

THE CARBON CYCLE

Climate change is a major challenge, but it's also an opportunity for us to reimagine what we can accomplish through agriculture. In addition to developing new solutions to reduce agriculture's impact on the environment, we're educating others on how to use agriculture as part of the solution. To help ensure a more sustainable future. We are taking steps toward a carbon-zero future: using natural affordable indigenous growing practices to reduce greenhouse gases from the atmosphere that the average farmer emits. Also, the development of climate-smart solutions including the GPO model and advancement in technology. GPO plant breeding techniques will help reduce agriculture's impact on climate change.

To help ensure a more sustainable future, farmers are taking steps toward a carbon-zero future: using cutting-edge tools and farming practices to remove as much—if not more—greenhouse gases from the atmosphere than a farmer emits. The development of climate-smart solutions including the GPO model and advancement in technology. GPO plant breeding techniques will help reduce agriculture's impact on climate change in the future.

The largest source of atmospheric carbon related to farming occurs when agricultural expansion leads to deforestation or draining of wetlands, which reduces the ability of the natural ecosystem to absorb and store carbon. By utilization of our developed solutions, this will help farmers grow enough on less land, agriculture is working to preserve natural habitats—even finding ways to help farmers identify areas of their land that would add more value to their operation as a biodiversity sanctuary to support animals, pollinators, and their surrounding environment to maintain a balanced ecosystem.

Effects on different elements air water & soil

On Earth, human activities are changing the natural greenhouse. Over the last century the burning of fossil fuels like coal and oil has increased the concentration of atmospheric carbon dioxide (CO_2). This happens because the coal or oil burning process combines carbon with oxygen in the air to make CO_2. To a lesser extent, the clearing of land for agriculture, industry, and other human activities has increased concentrations of greenhouse gases.

The consequences of changing the natural atmospheric greenhouse are difficult to predict, but certain effects seem likely:

- On average, Earth will become warmer. Some regions may welcome warmer temperatures, but others may not.
- Warmer conditions will lead to more evaporation and precipitation overall, but individual regions will vary, some becoming wetter and others dryer.
- A stronger greenhouse effect will warm the oceans and partially melt glaciers and other ice, increasing sea level. Ocean water also will expand if it warms, contributing further to sea level rise.

- Meanwhile, some crops and other plants may respond favorably to increased atmospheric CO_2, growing more vigorously and using water more efficiently. At the same time, higher temperatures and shifting climate patterns may change the areas where crops grow best and affect the makeup of natural plant communities.

THE SOLUTION
Hemp cannabis Sativa can and will reduce the carbon footprint of the world.

The Importance

When people talk about saving the environment, this movement is a direct response to the warming of Earth's climate. Global warming is a real threat to our farmers, our wildlife, and the world. The gas responsible for most of global warming is carbon dioxide, or CO2. Most of the CO2 released into the atmosphere comes from the combustion of fossil fuels for cars, factories, and electricity. Nitrous oxide is another toxic gas that is hurting our environment. Most nitrous oxide comes from fertilizers, industrial purposes, and the loss of our rainforests that would otherwise be storing CO2. While nitrous oxide is 300 times more harmful than carbon dioxide, no gas adds more warmth to the atmosphere than carbon dioxide.

When the topic of conversation moves to global warming, the conversation usually turns a little bleak. But there's new/old hope: hemp. Alright, maybe it's not new. Researchers claim that hemp was the first domesticated crop, back in 12,000 B.C. – and yet, it's new again. With the signing of the 2018 Farm Bill, hemp is now legal again in the US, opening the door for innovation and environmentally friendly alternatives.

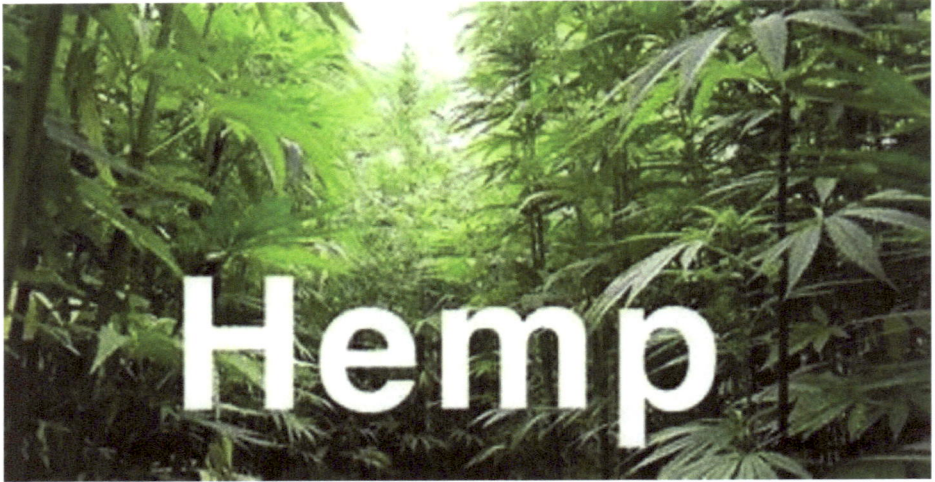

HEMP CANNABIS SATIVA

Hemp begins sequestering carbon the moment it is seeded; conservatively, hemp cultivation yields a sequestration ratio of about 1.5 units of sequestration per unit produced. One ton of harvested hemp fiber should sequester 1.62 tons of CO_2.

Hemp can also sequester carbon back into the soil through a process called biosequestration. In this process, hemp captures carbon emissions from the atmosphere. When the crop is harvested, the hemp can be slow-smoldered, not burned, to create biochar. This charcoal-based product can then be mixed with other nutrients and returned into the soil. Also, the hemp plant is a great rotation crop to revitalize the soil. It can remove radiation from the soil. According to a paper provided by Holon Ecosystem Consultants, hemp might give as much as 13 tons of charcoal per hectare annually, which would triple the output of Salix (a popular biomass crop) plantations.

HEMPCRETE

According to the United Nations Environment Program, the building sector contributes up to 30% of global annual greenhouse gas emissions and uses up to 40% of all energy. In recent years, China and Europe are neck and neck leading the way back into energy-efficient, hemp-based buildings, with Canada and the US following suit with hempcrete. Hempcrete is a fiber-reinforced material made from a combination of hemp hurd, lime, and water. This combination creates a bio fiber-reinforced material made from a combination of hemp hurd, lime, and water. This combination creates a bio-composite material that is lighter and more flexible than concrete while keeping the structural strength and thermal properties ideal for use in non-load bearing construction projects and insulation. But what about its carbon footprint? A life cycle analysis on a 120 square foot

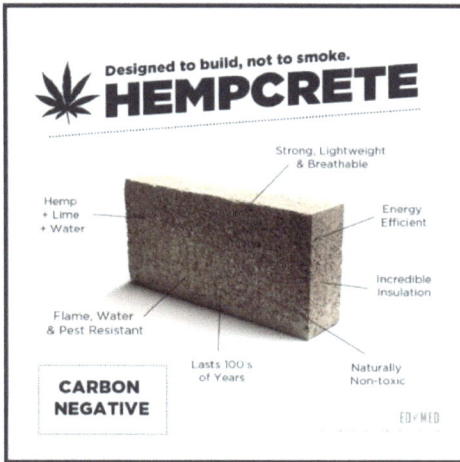

Hemp Hurd that can be used in Hempcrete

Hempcrete wall would find an emission of 3880 pounds of CO_2, but this would not include the materials' potential carbon sequestration. As we previously mentioned, one ton of hemp fiber should sequester 1.62 tons of CO_2 over the growing process, but hemp doesn't stop sequestering there. Over time, materials in hempcrete undergo calcination, absorbing more CO_2. Calcination is the process where materials heat up, below its boiling point, to drive off volatile matter or to effect changes. During this process, and under the same circumstances previously mentioned, the 120 square foot hempcrete wall should sequester 2400 pounds of CO_2, leaving only 1480 pounds of CO_2 created by the process. Those 1480 pounds are easily offset by the carbon sequestered in hemp production, leaving the entire project carbon negative.

HEMP BASED PLASTICS

Thermoplastics are a type of plastic made from polymer resins, like polyethylene, polypropylene, polystyrene, polycarbonate, and acrylic. These resins and natural materials can create anything from car parts to plastic ware. With this much potential, imagine the good that could be done by replacing a little plastic with hemp fibers. A study in 2003 by the University of Toronto found that we could save 50 000 MJ, or 3 tons of CO_2, per ton of thermoplastic by replacing 30% glass fiber reinforcement with 65% hemp fiber. We can also address the carbon storage capabilities in these natural fiber composites. The same writers of the University of Toronto's study estimated a carbon storage potential of 715 pounds per metric ton of hemp-based polypropylene composite. This means that the carbon is not sent back into the atmosphere but trapped in the material for years and years. Meaning the use of hemp for cars, homes, surfboards, planes, even weapons and technology can dramatically reduce the carbon output into the atmosphere.

GPO: GENETICALLY PURIFIED ORGANISMS INDIGENOUS GROWING PRACTICES

GPO is a Growing Pharms made company established by President Dj Jones Bey in order to specialize in the natural development of all pants and especially those used for human consumption. Such as cannabis in conjunction with the GP 885 exemption. This research and development uses ancient practices from DJ's Tribes along with his own research with Growing Pharms. He has been able to make dramatic advancement with plant genetics of all variations.

In GP we implement natural resources and high mineral based gems and stones to increase mineral levels in plants and food. By knowing the genetic makeup of a plant we know the minerals that are needed in order to produce the plant. In order to invoke a healthier mind, body and environment we only use elements given to use from the earth to increase plant quality. All the way from the medium to the water, ph. and mineral levels are all managed naturally with nothing from a factory. While also changing and reversing the negative effects of harmful chemicals that have altered plants over time products on the body and the world.

GROWING MEDIUMS

List Of Different Types Of Growing Mediums For Hydroponics

Coco Fiber   **Coco Chips**

"Coco coir" (Coconut fiber) is from the outer husk of coconuts. What was once considered a waste product, is one of the best growing mediums available. Although coco coir is an organic plant material, it breaks down and decomposes very slowly, so it won't provide any nutrients to the plants growing in it, making it perfect for hydroponics. Coco coir is also pH neutral, holds moisture very well, yet still allows for good aeration for the roots. Coco fiber comes in two forms, coco coir (fiber), and coco chips. They're both made of coconut husks, the only difference is the particle size. The coco fiber particle size is about the same as potting soil, while the coco chips particle size is more like small wood chips.

The larger size of the coco chips allows for bigger air pockets between particles, thus allowing even better aeration for the roots. Also, if you're using baskets to grow your plants in, the chips are too big to fall through the slats in the baskets. Both the fiber and chips come in compressed bricks, and once soaked in water it expands to about 6 times the original size. Coco fiber does tend to color the water, but that diminishes over time. And you can leach out most of the color if you soak it in warm/hot water a few times before use.

Composted and aged Pine bark

Pine bark is one of the first growing mediums used in hydroponics. It was generally considered a waste product, but has found uses as a ground mulch, as well as substrate for hydroponically grown crops. Pine bark is considered better than other types of tree bark because it resists decomposition better and has fewer organic acids that can leach into the nutrient solution than others. Bark is generally referred to as either fresh, composted, or aged.

Fresh bark uses up more nitrogen as it begins to decompose, so commercial growers generally compensate by adding extra nitrogen to the nutrient solution. During the composting process, nitrogen is added to the bark, and mixing it in while it breaks down. So, nitrogen issues are far less of a concern with composted pine bark. Ageing is a similar process, but has less nitrogen added to it, so it's better than using fresh bark, but not as easy as the composted bark. Pine bark can be found at places that sell ground mulch, as well as ground mulch for playgrounds.

Grow Rock (Hydro corn)

Grow rock is a Lightweight Expanded Clay Aggregate (L.E.C.A.), That is a type of clay which is super-fired to create a porous texture. It's heavy enough to provide secure support for your plant's, but still light weight. Grow rocks are a non-degradable, sterile growing medium that holds moisture, has a neutral pH, and will pick up nutrient solutions to the root systems of your plants. Hydro corn grow media is reusable, it can be cleaned, sterilized, then reused again. Although on a large scale, cleaning and sterilizing large amounts of grow rocks can be quite time consuming. Grow rock is one of the most popular growing mediums used for hydroponics, and just about every store selling hydroponics supplies carries it.

Growstone Hydroponic Substrate

Growstones are made from recycled glass. They grow like rocks (hydro corn) but are made of clay and shaped marbles. Growstones are lightweight, unevenly shaped, porous, and reusable, they provide good aeration and moisture to the root zone. They have good wicking ability and can wick water up to 4 inches above the water line. So, you'll want to make sure it has good drainage or is deep enough, so it doesn't wick water all the way to the top. Otherwise like with the growing media in any hydroponic system, if the top of the growing media is continually wet, you may have problems with stem rot. While they are made from recycled glass, they're not sharp and you won't get cut from it, even if they break.

Perlite

Perlite is mainly composed of minerals that are subjected to very high heat, which then expand it like popcorn, so it becomes very light weight, porous and absorbent. Perlite has a neutral pH, excellent wicking action, and is very porous. Perlite can be used by itself or mixed with other types of growing mediums. However, because perlite is so light that it floats, depending on how you designed your hydroponic system, perlite by itself may not be the best choice of growing media for flood and drain systems.

Perlite is widely used in potting soils, and any nursery should carry bags of it. However, perlite is sometimes also used as an additive added to cement. You may find it for a better price with the building supply's, and/or at places that sell concrete mixes and mixing supplies. When working with perlite be careful not to get any of the dust in your eyes. Rinse it off to wash out the dust and wet it down before working with it to keep the dust from going airborne.

Pine shavings

Pine shavings are an inexpensive hydroponic growing media as well, and a lot of commercial growers use it. Generally, for large scale hydroponic drip irrigation systems. Don't confuse pine shavings with saw dust. Saw dust will become compact and waterlogged easily. You'll want to make sure your pine shavings were made from kiln dried wood and does "NOT" contain any chemical fungicides. Kiln dried to burn off all the sap in the wood that is bad for the plants. Most pine shaving products would be kiln dried to begin with.

Good source to find pine shavings are pet supply stores. It's used for things like hamsters and rabbit bedding. Just make sure to read the package to be sure it doesn't have any chemical additives like fungicides or odor inhibitors. You should be fine if it states it's organic. Another good cheap source for pine shavings is at feed stores, it's also used as bedding in horse stalls and they sell it by cubic yard. If you have a choice to get the largest particle size, you can. The larger the air pockets between the shavings, the better aeration to your roots.

Pine shavings are a wood product, so they absorb water easily, thus can become waterlogged easily. So, make sure you have good drainage, so the shavings don't sit in water. If there is a possibility of it sitting in water, a layer of rocks at the bottom will aid drainage greatly.

Rice Hulls

Depending on your location, rice hulls may be readily available. It's a byproduct of the rice industry. Even though they are an organic plant material, they break down very slowly like coco coir, making them suitable as a growing media for hydroponics. Rice hulls are referred to as either fresh, aged, composted and parboiled, or carbonized. Fresh rice hulls are typically avoided as a hydroponic growing media because of the high probability of contaminants such as rice, fungal spores, bacteria, decaying bugs, and weed seeds. Parboiled rice hulls (PRH) is done by stemming and drying the rice hulls after the rice has been milled from them. This kills any spores, bacteria, and microorganisms, leaving a sterile and clean product.

Rice hulls are also often used as part of a mix of growing media such as 30%-40% rice hulls and pine bark mix. The overall pH of parboiled and composted rice hulls ranges from 5.7 to 6.5, which is right in the pH range for most hydroponically grown plants. Fresh and/or composted rice hulls tend to have a high Manganese (Mn) content. But problems with Manganese toxicity can be avoided if the pH is above 5. Which is below normal range for hydroponics anyway?

River rock

River rock is common and easy to find in home improvement stores, as well as even pet supply's stores (with the fish and aquariums). River rock is inexpensive (depending on where you get it from) and comes in many different sizes. River rock is rounded with smooth edges from tumbling down the river. Though manufactured river rock is rounded using large mechanical tumblers, it has the same result with smooth edges.

You can use regular rocks from your backyard in hydroponic systems as well if you don't mind the jagged edges. Just make sure to clean and sanitize them before using it. Just spray all the dirt off the rock using the jet spray from your hose to clean it, then soak it overnight in bleach water to sanitize it. Then just rinse and use. Though using rock as a growing media is inexpensive and easy, it will get heavy quickly, so you won't want to move it later.

River rocks are not porous, therefore it doesn't hold and retain moisture in the root zone of hydroponic systems. Rock is uneven so it has a lot of air pockets between the rocks so the roots can get plenty of oxygen, but water easily drains down to the bottom. Rock won't wick up moisture either, so you will need to adjust your watering schedules, so the roots don't dry out between waterings. You can mix in some coco chips or other growing media that holds moisture with your rock to aid it in holding onto moisture longer.

Because of the good drainage properties of rock, it's very good to use to aid in the drainage of other hydroponic growing media that might otherwise become saturated from sitting in water. A layer of rock at the bottom of the growing container will keep your growing media from sitting in water at the bottom of the container, keeping it from being saturated.

Rockwool

Rockwool is one of the most common growing mediums used in hydroponics. Rockwool is a sterile, porous, non-degradable medium that is composed primarily of granite and/or limestone which is super-heated and melted, then spun into a small thread like cotton candy. The rockwool is then formed into blocks, sheets, cubes, slabs, or flocking. Rockwool sucks up water easily so you'll want to be careful not to let it become saturated, or it could suffocate your plants roots, as well as lead to stem rot and root rot. Rockwool should be pH balanced before use. That's done by soaking it in pH balanced water before use.

Sand

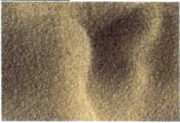

Sand Is actually a very common growing media used in hydroponics. It's the main growing media used at the Epcot Center Hydroponic Greenhouse in Florida. Mainly for their large hydroponically grown plants and trees. Sand is like rock, just smaller in size. Because the particle size is smaller than regular rock, moisture doesn't drain out as fast. Sand is also commonly mixed with Vermiculite, Perlite, and or coco coir. All help retain moisture as well as help aerate the mix for the roots.

When using sand as a growing media you will want to use the largest grain size you can get. That will help increase aeration to the roots by increasing the size of the air pockets between the grains of sand. Mixing Vermiculite, Perlite, and or coco coir with the sand will also help aerate. You will also want to rinse the sand well before use to get as much of the dust particles out of it as you can. One big downside to using sand as a growing media for hydroponics is that it is very heavy. 3-4 gallons of wet sand can weigh up to 50lbs. So, you won't want to be moving it once you get it set up. Or use it in a ratio of something like 20%-30% sand and the rest Vermiculite, Perlite, or another type of growing media to reduce weight.

Vermiculite

Vermiculite is a silicate mineral that, like perlite, expands when exposed to very high heat. As a growing media, vermiculite is quite like perlite except that it has a relatively high cation-exchange capacity, meaning it can hold nutrients for later use. Also like the perlite, vermiculite is very light and tends to float. There are different uses and types of vermiculite, so you'll want to be sure what you get is intended for horticulture use. The easiest way to be sure is to get it from a nursery.

Water absorbing crystals (water-absorbing polymers)

Water absorbing polymer crystals have been around for quite a while and are used in many industries. Everything from baby diapers, to the sports industry where they are used in cloth rags they can wear on the head or neck to keep cool. They are also used in gardening where the crystals are mixed into the soil to help retain moisture in the soil. Florists use them in vases to keep flowers fresh, and the colored ones make for a nice decorated display.

The crystals expand to many times their size as they soak up water. One pound of the crystals can hold as much as 50 gallons of water. The crystals come in many sizes, everything from a powder, to marble and even golf ball size. Depending on the size of the crystals they can take more than an hour or two to fully absorb. When they are full of water they look and feel like a glob of jelly. Once they dry out, they can be stored and reused again over and over.

The water absorbing polymer crystals are not a common hydroponic growing media, but like everything else, it's growing in popularity. Mostly due to their increased availability. They are quite inexpensive, and reusable. However, used alone by themselves they don't allow the roots to get much oxygen/air. Being like jelly they pack together and fill the air pockets. The larger size crystals are better suited for use in hydroponics. The larger size helps retain some of the air pockets between the crystals. Also, by mixing some river rock or other similar growing media with the crystals will help increase the air pockets between the crystals.

Using the polymer crystals for hydroponics allows for some of the simplest hydroponic system designs. Even on the slimmest of budgets. Simply soaking some water absorbing crystals in a nutrient solution, then setting them in a container and placing your seedling's in it, you've got a hydroponically grown plant. You don't need any pumps. Just make sure there are holes in the bottom of your container, and just place your container in nutrient solution once or twice a week to rehydrate the crystals.

You won't find water absorbing polymer crystals in hydroponic shops, but they are easy to find. Because of their popularity, most large nurseries carry them as soil amendments. Also, if you do a search for them online, you'll find hundreds of places selling them.

GEM AND STONES

This list of stones can be mixed to match the needs of each plant, like how cannabis is composed of minerals Also stones and gems can be mixed into other mediums

Air, water, and light; you get those from the environment in which you grow your plants. But of course, there are many hugely important nutrients in the soil you're using to grow your cannabis. Of the roughly 16 essential nutrients that support virtually every plant in existence, there are three that are the most important. Your plants need these three nutrients in larger quantities than the rest. Nitrogen (N), Phosphorus (P), and Potassium (K). N-P-K. As a rule, go strong on the phosphorus and potassium, and not as much on the nitrogen, you will have an overall healthy cannabis plant.

Nitrogen (N)

If your soil is nutrient-deficient, it's likely nitrogen. Cannabis uses nitrogen to develop. Nitrogen forms the building blocks for the plant's proteins, its chlorophyll, its enzymes, basically everything that makes the plant function. Nitrogen helps cannabis create cannabinoids. You end up overfeeding the nutrient to your plants, and that will slow its root growth and its flowering. Nitrogen makes your plants "mature," and you don't want them to grow up too fast.

Phosphorus (P)

Phosphorus is what makes it flower. It's what brings the bloom phase. Phosphorus helps the plant produce the enzymes and nucleic acids that go into flowers. If your plants reach maturity with low phosphorus, you'll see smaller flowers and end up with less product. Phosphorus deficiency is what turns cannabis plants purple, a buildup of anthocyanins.

Potassium (K)

Potassium is the nutrient that gives your plants their strength. It accounts for around one percent of the total dry weight of cannabis plants. Your plants need potassium to breathe and make proteins. They also need it to keep up the internal pressure that helps your plants stand tall and strong. Deficiencies in this important nutrient will reveal itself through dead spots on leaves, flimsy stalks, and problems falling over. Low K also leaves the roots more vulnerable to disease.

Calcium

Plants need calcium for a different reason., Most of the calcium in cannabis plants is used for defense. The calcium acts as a shield against other toxic elements that can poison and kill your plants. There are lots of toxins in the soil with similar structures to calcium that try to imitate it. Calcium deficiency will make the leaves of your plants twist up and deformed. So, use dolomite limestone and other calcium-based stones to make sure your soil is well supplied with calcium before you plant.

Sulfur

Even though it never stays around in the final product, Sulfur is essential for healthy cannabis plants. All the proteins made by the cannabis plant are built of two indispensable amino acids. Sulfur is a very common soil pollutant, too much will kill the plant's chlorophyll, which it needs to turn light into food. Calcium protects the plants from the excess sulfur. Cannabis plants can benefit from it at key stages in the grow process.

Important Micronutrients for Healthy Cannabis Plants

The above nutrients are what makeup virtually all the tissue of your cannabis plants. But there's still 0.5 percent of that tissue which is composed of "micronutrients" found in soil.

They are:

- **Iron**
- **Molybdenum**
- **Zinc**

- **Manganese**
- **Chlorine**
- **Magnesium**
- **Copper**
- **Boron**

like iron and boron. All these metal nutrients are toxic at anything higher than trace levels, and your plants don't need more than very small amounts. It's likely you have too many of them in your soil than too little. Signs that you have too much of these elements include yellowing, dead, or deformed leaves. If you think you may have too little of these trace nutrients. Add some chelating agents to your soil. These convert the metals into soluble versions (called "salts," FYI) that roots can absorb.

Alphabetical List of Precious and Semiprecious Gemstones to mix as Grow Medium

A gemstone is a crystalline mineral that can be cut and polished to make jewelry and other ornaments. The ancient INDIGENOUS PEOPLES OF America (the so called Indian) and Indigenous people of Africa made a distinction between precious and semiprecious gems, which is still used. Precious stones were hard, rare, and valuable. The only "precious" gemstones are diamond, ruby, sapphire, and emerald. All other quality stones are called "semiprecious," even though they may not be any less valuable or beautiful. Today, mineralogists and gemologists describe stones in technical terms, including their chemical composition, Mohs hardness, and crystal structure

Agate

Agate is a striped or banded form of the mineral chalcedony.

Agate is cryptocrystalline silica, with a chemical formula of SiO_2. It is characterized by rhombohedral microcrystals and has a Mohs hardness ranging from 6.5 to 7. Chalcedony is one example of gemstone quality agate. Onyx and banded agate are other examples.

Agate is an excellent stone for rebalancing and harmonizing body, mind and spirit. It cleanses and stabilizes the aura, eliminating and transforming negativity. Agate enhances mental function, improving concentration, perception and analytical abilities. It soothes and calms, healing inner anger or tension and creates a sense of security and safety. Agate heals the eyes, stomach and uterus; cleanses the lymphatic system and the pancreas; strengthens blood vessels and heals skin disorders.

Alexandrite or Chrysoberyl

Chrysoberyl is a gemstone made of beryllium aluminate. Its chemical formula is $BeAl_2O_4$. Chrysoberyl belongs to the orthorhombic crystal system and has a Mohs hardness of 8.5. Alexandrite is a strongly pleochroic form of the gem that can appear green, red, or orange yellow, depending on how it is viewed in polarized light.

Chrysoberyl is an effective protective stone and, since ancient times, has been used to keep disaster at bay. Chrysoberyl transforms negative thoughts into positive energy. It increases self-confidence and strengthens self-worth. Chrysoberyl helps

you to see both sides of a situation or problem. It brings compassion and generosity and encourages forgiveness. Chrysoberyl brings the qualities of discipline and self-control. It promotes concentration and the ability to learn, enabling the wearer to think clearly and farsighted. Chrysoberyl aligns the solar plexus and crown chakras. It opens the crown chakra and increases both spiritual and personal power. Chrysoberyl is associated with wealth and is excellent for creativity. It also promotes tolerance and harmony.

Chrysoberyl helps to highlight the cause of dis-ease. It supports self-healing. Chrysoberyl balances adrenaline and cholesterol and fortifies the chest and liver.

Amber

Gemstone-quality amber is translucent. 97

Although amber is considered a gemstone, it's organic rather than an inorganic mineral. Amber is fossilized tree resin. It's usually golden or brown and may contain inclusions of plants or small animals. It is soft, has interesting electrical properties, and is fluorescent. Generally, the chemical formula of amber consists of repeating isoprene (C_5H_8) units.

Amber is a powerful healer and cleanser of the body, mind and spirit. It also cleanses the environment. Amber draws disease from the body, healing and renewing the nervous system and balancing the right and left parts of the brain. It absorbs pain and negative energy, helping to alleviate stress. Amber clears depression stimulates the intellect and promotes self-confidence and creative self-expression. It encourages decision-making, spontaneity and brings wisdom, balance and patience.

Amber opens the throat center, treating goiters and other throat problems. It also treats stomach, spleen, kidneys, bladder, liver and gallbladder. Amber strengthens the mucus membranes and alleviates joint problems.

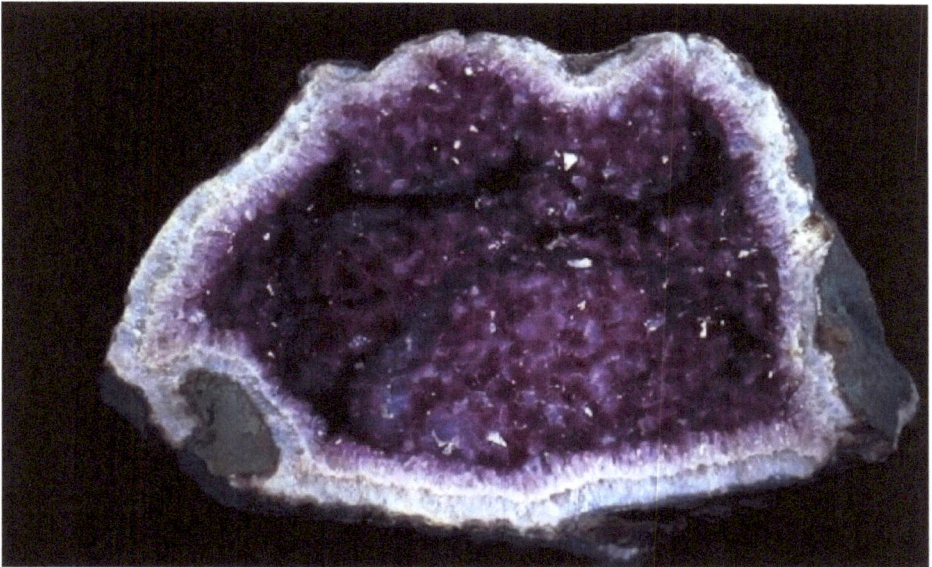

Amethyst

Amethyst gemstone is a purple variety quartz.

Amethyst is a purple variety of quartz, which is silica or silicon dioxide, with a chemical formula of SiO_2. The violet color comes from irradiation of iron impurities in the matrix. It is moderately hard, with a Mohs scale hardness of around 7.

Amethyst is a powerful and protective stone. It guards against psychic attack, transmuting the energy into love and protecting the wearer from all types of harm, including geopathic or electromagnetic stress and ill wishes from others. Amethyst is a natural tranquilizer, it relieves stress and strain, soothes irritability, balances mood swings, dispels anger, rage, fear and anxiety. Alleviates sadness and grief, and dissolves negativity. Amethyst activates spiritual awareness, opens intuition and enhances psychic abilities. It has strong healing and cleansing powers. Amethyst encourages sobriety, having a sobering effect on overindulgence of alcohol, drugs or other addictions. It calms and stimulates the mind, helping you become more focused, enhancing memory and improving motivation. Amethyst assists in remembering and understanding dreams. It relieves insomnia. Encourages selflessness and spiritual wisdom.

Amethyst boosts hormone production tunes the endocrine system and metabolism. It strengthens the immune system, reduces pain and strengthens the body to fight against cancer. It destroys malignant tumors and aids in tissue regeneration. Cleanses the blood. Relieves physical, emotional and psychological pain or stress. Amethyst eases headaches and releases tension. It reduces bruising, swellings, injuries, and treats hearing disorders. Amethyst heals diseases of the lungs and respiratory tract, skin conditions, cellular disorders and diseases of the digestive tract.

Apatite

Apatite is a relatively soft blue-green gem. Apatite is a phosphate mineral with the chemical formula $Ca_5(PO_4)_3(F, Cl, OH)$. It's the same mineral that comprises human teeth. The gemstone form of the mineral displays the hexagonal crystal system. Gems may be transparent or green or less commonly other colors. It has a Mohs hardness of 5.

Apatite is a stone of manifestation. It is related to service and to humanitarian pursuits. Apatite is attuned to the future yet connects to past lives. It stimulates the development of our psychic gifts and spiritual attunement, deepening meditation and aiding communication and self-expression on all levels. It balances the physical, emotional, mental and spiritual bodies, and the chakras, eliminating overactivity and stimulating underactivity. Psychologically, Apatite increases motivation. It draws off negativity about oneself and others. It is helpful for hyperactivity and autism in children. Apatite enhances creativity and the intellect. It clears confusion and frustration, reducing irritability and awakening the inner self. Apatite expands knowledge and truth and eases sorrow, apathy, and anger.

Physically, Apatite aids in the absorption of calcium, helping cartilage, bones and teeth; healing bones and encouraging the formation of new cells. It improves arthritis and joint problems. Apatite can successfully help in suppressing hunger and raising the metabolic rate, encouraging healthy eating. It heals the glands, meridians, organs and overcomes hypertension.

Diamond

Pure diamond is colorless crystal carbon with a high refractive index. Diamond is pure carbon in a cubic crystal lattice. Because it's carbon, its chemical formula is simply C (the element symbol of carbon). Its crystal habit is octahedral and it is extremely hard (10 on the Mohs scale). This makes diamond the hardest pure element. Pure diamond is colorless, but impurities produce diamonds that may be blue, brown, or other colors. Impurities may also make diamond fluorescent.

A symbol of purity, the Diamond's pure white light helps to bring our lives into a cohesive whole. It brings love and clarity into partnerships, bonding relationships. Diamond is a sign of commitment and fidelity and instills trust to relationships and situations. It inspires the forces of accumulation, attracting the manifestation of abundance. Diamond is an energy amplifier. It is a stone that never requires recharging. It will bring strength and endurance to all energies and will enhance the power of other crystals. However, beware, as this means it will increase negative energy as well as positive! Excellent for blocking electromagnetic stress and for protection against cell phone emanations. Diamond imparts fearlessness, invincibility and fortitude. It clears emotional and mental pain, reducing fear and bringing about new beginnings. Stimulates creativity, inventiveness, imagination and ingenuity. It brings clarity of mind and aids enlightenment. Diamond allows the soul light to shine out. It aids spiritual evolution and reminds you of your soul's aspirations.

Diamond purifies and detoxifies all the body's systems, rebalancing the metabolism, and building up stamina, strength and treating allergies and chronic conditions. It also helps glaucoma and clears sight. Diamond effectively treats dizziness and vertigo and benefits the brain. Counteracts poisoning.

Emerald
The green gemstone form of beryl is called emerald.

Emerald is the green gemstone form of the mineral beryl. It has a chemical formula of $(Be_3Al_2(SiO_3)_6)$. Emerald displays a hexagonal crystal structure. It is very hard, with a rating of 7.5 to 8 on the Mohs scale.

Emerald is known as the "stone of successful love". It brings loyalty and provides for domestic bliss. It enhances unconditional love, unity and promotes friendship. Keeps partnerships in balance and can signal unfaithfulness if it changes color. Emerald stimulates the heart chakra, having a healing effect on the emotions as well as the physical heart. It ensures physical, emotional and mental equilibrium, bringing harmony to all areas of one's life. Focusing intention and raising consciousness, it brings in positive actions, eliminating negativity and enhancing the ability to enjoy life to the fullest.
Emerald enhances psychic abilities, opens clairvoyance, and stimulates the use of greater mental capacity. It helps bring

awareness of the unknown to conscious recognition, imparting reason and wisdom. Emerald assists in inciting activity and focus in one's actions. It strengthens memory and imparts clarity of thought. It inspires a deep inner knowing, promoting truth and discernment.

Emerald treats disorders of the heart, lungs, spine and muscular system. It aids in recovery after infectious illness, helps sinuses and soothes the eyes, improving vision. It has a detoxifying effect on the liver and alleviates diabetes and rheumatism.

Garnet

Grossular var. Hessonite. Garnet comes in several colors and crystal forms.

Garnet describes any member of a large class of silicate mineral. Their chemical composition varies but may be generally described as $X_3Y_2(\text{SiO}_4)_3$. The X and Y locations may be occupied by a variety of elements, such as aluminum and calcium. Garnet occurs in almost all colors, but blue is extremely rare. Its crystal structure may be a cubic or rhombic dodecahedron, belonging to the isometric crystal system. Garnet ranges from 6.5 to 7.5 on the Mohs scale of hardness. Examples of different types of garnets include pyrope, almandine, spessartine, hessonite, tsavorite, uvarovite, and andradite.

Garnets are not traditionally considered precious gems, yet a tsavorite garnet may be even more expensive than a good emerald.

Garnet cleanses and re-energizes the chakras. It revitalizes, purifies and balances energy, bringing serenity or passion as appropriate. Inspires love and devotion. Garnet balances the sex drive and alleviates emotional disharmony. It activates and strengthens the survival instinct, bringing courage and hope. Stimulates past-life recall. Sharpens perceptions of oneself and others. Garnet removes inhibitions and taboos. It opens the heart and bestows self-confidence.

Garnet regenerates the body and stimulates the metabolism. It treats disorders of the spine and spinal fluid, bone, cellular structure and composition. Purifies the heart, lungs, blood and regenerates DNA. Garnet boosts the immune system and energy levels.

Opal

Opal is a soft silicate gemstone. Opal is hydrated amorphous silica, with the chemical formula ($SiO_2 \cdot nH_2O$). It may contain from 3% to 21% water by weight. Opal is classified as a mineraloid rather than mineral. The internal structure causes the gemstone to diffract light, potentially producing a rainbow of colors. Opal is softer than crystal silica, with a hardness of around 5.5 to 6. Opal is amorphous, so it does not have a crystal structure.

Opal is an emotional stone and reflects the mood of the wearer. It intensifies emotions and releases inhibitions. Encourages both freedom and independence. Opal enhances cosmic consciousness and induces psychic and mystical visions. It stimulates originality and creativity. Helps to release anger and claim self-worth, aiding in accessing and expressing one's true self. Opal strengthens memory. It encourages an interest in the arts. Wearing Opal brings loyalty, faithfulness and spontaneity.

Opal strengthens the will to live and treats infections and fevers. Purifying the blood and kidneys, Opal also regulates insulin. It eases childbirth and alleviates PMS.

Pearl

Pearl is an organic gemstone produced by a mollusk. Like amber, a pearl is an organic material and not a mineral. Pearl is produced by the tissue of a mollusk. Chemically, it is calcium carbonate, $CaCO_3$. It is soft, with a hardness of around 2.5 to 4.5 on the Mohs scale. Some types of pearls display fluorescence when exposed to ultraviolet light, but many do not.

Pearl signifies faith, charity and innocence. It enhances personal integrity and helps to provide a focus to one's attention. Pearl symbolizes purity and is known as a "stone of sincerity". It brings truth to situations and loyalty to a "cause". Inhibits boisterous behavior.

Pearl treats digestive disorders and the soft organs of the body. It relieves conditions of bloating and biliousness. Pearl increases fertility and eases childbirth.

Peridot

Peridot is a green gemstone. Peridot is the name given to gem-quality olivine, which has the chemical formula $(Mg, Fe)_2SiO_4$. This green silicate mineral gets its color from magnesium. While most gems occur in different colors, peridot is found only in shades of green. It has a Mohs hardness of around 6.5 to 7 and belongs to the orthorhombic crystal system.

Peridot is a powerful cleanser. It releases and neutralizes toxins on all levels. Alleviates jealousy, resentment, spite, bitterness, irritation, hatred and greed. Reduces stress, anger and guilt. Peridot opens our hearts to joy and new relationships. It enhances confidence and assertion, motivating growth and change. Sharpens and opens the mind to new levels of awareness. Banishes lethargy, apathy and exhaustion. Peridot enables you to take responsibility for your own life.

Peridot strengthens the immune system, metabolism and benefits the skin. It aids disorders of the heart, thymus, lungs, gallbladder, spleen and intestinal tract. Treats ulcer and strengthens eyes. Balances bipolar disorders and overcomes hypochondria.

Quartz

Quartz is a silicate mineral with the repeating chemical formula SiO_2. It may be found in either the trigonal or hexagonal crystal system. Colors range from colorless to black. Its Mohs hardness is around 7. Translucent gemstone-quality quartz may be named by its color, which it owes to various element impurities. Common forms of quartz gemstone include rose quartz (pink), amethyst (purple), and citrine (golden). Pure quartz is also known as rock crystal.

Clear Quartz is known as the "master healer" and will amplify energy and thought, as well as the effect of other crystals. It absorbs, stores, releases and regulates energy. Clear Quartz draws off negative energy of all kinds, neutralizing background radiation, including electromagnetic smog or petrochemical emanations. It balances and revitalizes the physical, mental, emotional and spiritual planes. Cleanses and enhances the organs and subtle bodies and acts as a deep soul cleanser, connecting the physical dimension with the mind. Clear Quartz enhances psychic abilities. It aids concentration and unlocks memory. Stimulates the immune system and brings the body into balance. Clear Quartz (Crystal Quartz, Rock Crystal) harmonizes all the chakras and aligns the subtle bodies.

Ruby

Ruby is the red gemstone form of the mineral corundum. Pink to red gemstone-quality corundum is called ruby. Its chemical formula is Al_2O_3Cr. The chromium gives ruby its color. Ruby exhibits a trigonal crystal system and a Mohs hardness of 9.

Ruby encourages passion and a zest for life. It improves motivation and setting of realistic goals. Balances the heart and instills confidence. Ruby encourages joy, spontaneity, laughter and courage. It promotes positive dreams and stimulates the pineal gland. Aids in retaining wealth and passion. Ruby encourages removal of negative energies from your path. It overcomes exhaustion and lethargy and imparts potency and vigor. Calms hyperactivity.

Ruby detoxifies the body, blood and lymphatic system. It treats fevers, infectious disease and restricted blood flow. Ruby stimulates the adrenals, kidneys, reproductive organs and spleen.

Sapphire

Sapphire is any gemstone-quality corundum that is not red.

Sapphire is any gem-quality specimen of the aluminum oxide mineral corundum which is not red. While sapphires are often blue, they can be colorless to any other color. Colors are due to trace amounts of iron, copper, titanium, chromium, or magnesium. The chemical formula of sapphire is (α-Al_2O_3). Its crystal system is trigonal. Corundum is hard, around 9 on the Mohs scale.

Known as the "wisdom stone", each color of Sapphire brings

its own wisdom. It releases mental tension, depression, unwanted thoughts and spiritual confusion. Sapphire restores balance within the body, aligning the physical, mental and spiritual planes, bringing serenity and peace of mind. It stimulates concentration, brings lightness, and joy. Sapphire is also known as a "stone of prosperity", attracting gifts of all kinds and fulfilling dreams and desires.

Sapphire treats blood disorders, combating excessive bleeding and strengthening the walls of the veins. It is used for cellular disorders, regulates the glands and calms overactive body systems.

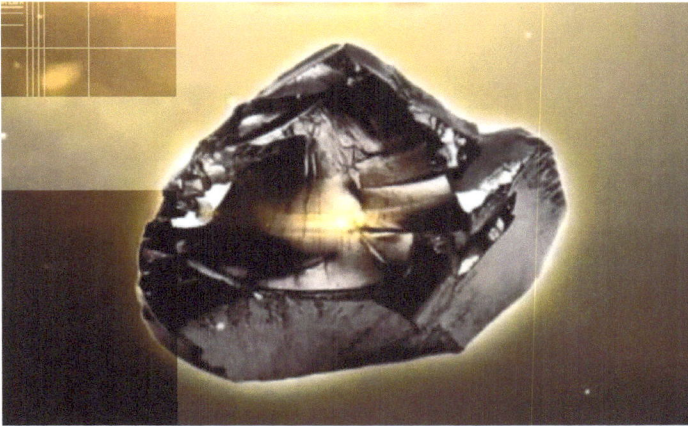

SHUNGITE

Surprisingly, shungite is a mineral that has yet to garner the same popularity in the crystal healing community as stones of a similar dark luster like black tourmaline. While not a crystal, shungite is one of the strongest mineral healers. It is the only mineral that has been scientifically proven to possess "fullerenes," molecules with powerful healing qualities. With a Mohs score of 4, it is a dense rock. Its real strength comes when tested by fire. Shungite's ability to withstand fire and keep from melting has given it the classification of a pyrobitumen. Shungite's surface luster ranges from semi-metallic, to matte black. Inside, shungite is a noncrystalline carbon mineraloid.

Shungite in agriculture is used in both crops and livestock. The plant uses shungite-dolomite fertilizer that reduces the acidity of the soil and helps to preserve moisture in the soil at 2-2.5 times longer than in areas without shungite, and has a positive effect on the productivity and efficiency of one-year agrocenosis potatoes. The effect on the yield and quality of vetch-oat mixture shungite-dolomite fertilizer does not yield complete fertilizer. It is more than a complete fertilizer, increases the content of phosphorus and calcium in plant mass and increases the total collection of digestible protein. Application shungite productivity increases potatoes by 70% and improves the sustainability of the tubers to a complex disease.

SHUNGITES-dolomite dressing contains:

- Calcium and magnesium - essential for plant growth.

- Trace elements: potassium, phosphorus, vanadium, cobalt, copper, nickel, zinc, etc., as well as rare earth, accelerating the growth of plants.

Black coloring grains shungite regulates the thermal regime of the soil: accumulation, preservation and long slow return absorbed per day of solar heat. Fertilizing loosens the soil, due to the presence in it of grains shungite 1 - 3 mm.

Applicable for all kinds of vegetables, fruit trees and bushes. Increases productivity by 1.5 - 2 times.

Application rate: 300 - 400 g / m2 - for light soils (sandy, light loam), 400 - 600 g/m2 - for heavy soils (peat, medium and heavy).

We also investigated the effect on the growth of the seeds when watering shungite water. Studies have shown that watering seed shungite water can speed up germination, growth and development. Adding 10 g. per 1 m2 of soil can provide a lot of mineral plant nutrients. In the soil houseplants can make 10 grams of flour shungite 1 kg of substrate. Crushed stone used as a drainage and to feed the pot and garden plants.

Using shungite is economically very profitable and does not require large capital expenditures.

Shungite healing properties span the board from purity to protection. Shungite has electricity conducting properties. These shungite properties are also known to aid in the inhibiting of EMFs, or electromagnetic fields, that are the result of electromagnetic radiation. Electromagnetic fields are created by electronic devices. Common electronic items such as laptops, cell phones, computers and tablets all put out EMFs. Placing shungite at the base of a computer, microwave or around your various home electronic devices will not interfere with their operations but will block out some of their free radial output. It's also used as an elixir. Shungite elixirs can be made by simply placing the mineral in water and allowing it a short time to purify. Drinking this purified water is said to have many beneficial qualities such as an increased rate of healing and cell growth, and detoxification of the body. Holding and meditating amplifies the shungite healing properties of headaches, back pain, blood pressure, inflammation and muscle pain. The antioxidants in shungite are forceful impacts on health and a fully functioning immune system. Shungite healing properties are powerful for the spirit as well. It can be used to combat insomnia, boost energy, reduce stress and relieve anxiety.

Shungite cleans water and neutralizes impurities. It is highly absorbent, enough to draw contaminants from water. Also, when placed in water, fullerenes attract and eliminate waterborne contaminants.

A shungite water purifier cleans water by removing:

- Phenol
- Iron
- Manganese
- Chlorine

- Nitrates
- Nitrites
- Bacteria
- Microorganisms

When infused with shungite, water becomes acidic, so it requires alkalization before it can be consumed in large quantities. The ideal pH standard for drinking water is 7.2 to 7.3, whereas shungite water pH ranges from 3 to 5.5. Consuming large amounts of untreated shungite water can cause health problems like gastritis.

The Russians typically use dolomite and quartz sand to reduce the acidity of shungite water. The acidity dissolves the dolomite which, in turn, alkalizes the water and enriches it with calcium, magnesium, and other minerals.

The quartz sand then traps heavy metals and radionuclides, helping to restore the structure of the water molecules.

Our filters work in a similar way to provide you with pure, drinkable shungite water. Contact High Water Standard today to get a shungite water filter installation that can provide you with instant access to clean, healthy water.

Health-related shungite water benefits

While drinking clean, fresh water can benefit your health in many ways, Russians consider shungite water to be especially revitalizing because of the fullerenes it contains.

Shungite water benefits the human body by normalizing cellular metabolism and increasing enzyme activity. Fullerenes boost the regenerative ability of human tissue and influence the exchange of neurotransmitters, which can improve a person's resistance to stress.

Fullerenes also have antihistamine and anti-inflammatory properties, helping to relieve pain and suppress allergy symptoms. Shungite water science shows that fullerenes are effective even in very small doses and the benefits can last for months.

Other shungite water benefits include its antioxidant properties, which help to suppress free radicals. Many believe the antioxidant ability of fullerenes to exceed that of vitamin E, dibunol, and beta-carotene.

Some people living with psoriasis find that dabbing the skin with shungite water a couple of times a day is highly effective at treating their condition.

Drinking shungite water may also be beneficial in the treatment of:

- Cardiovascular diseases
- Kidney diseases
- Liver diseases
- Pancreatic diseases
- Cholecystitis
- Gastritis and dyspepsia
- Gall-bladder problems
- Anemia
- Asthma
- Chronic fatigue
- Arthritis

Because of its quality and purity, shungite water is an excellent addition to any detox program. Athletes and gym-goers can also drink shungite water after a workout to help revitalize fatigued muscles.

You can get your own shungite water purification system at High Water Standard.

Reference profile

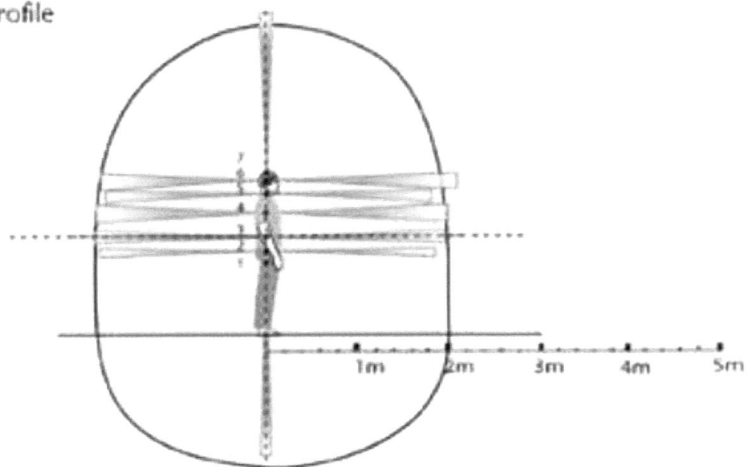

Profile with shungite

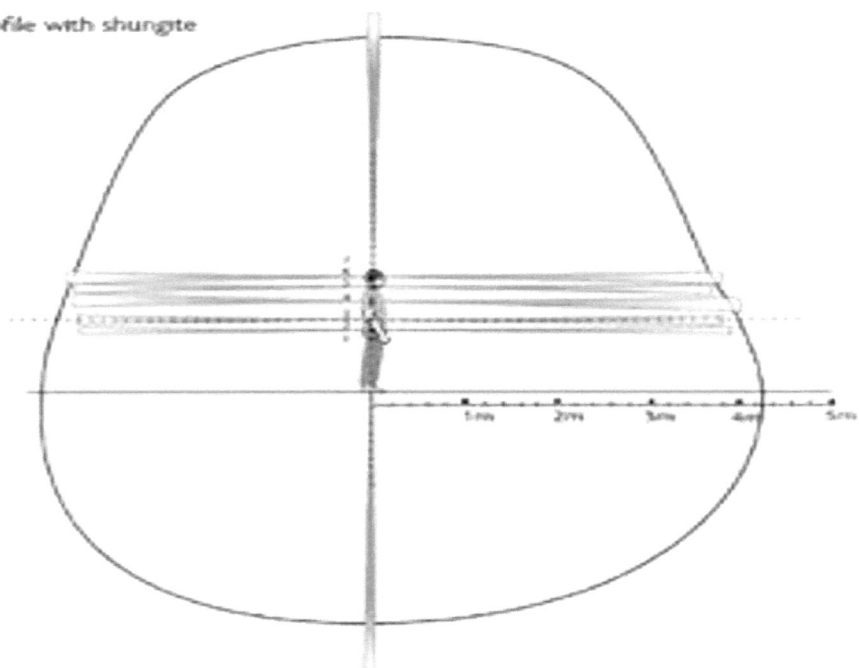

Topaz

Topaz is a silicate gemstone that occurs in many colors.

Topaz is a silicate mineral with the chemical formula $Al_2SiO_4(F, OH)_2$. It belongs to the orthorhombic crystal system and has a Mohs hardness of 8. Topaz may be colorless or nearly any color, depending on impurities.

Topaz soothes, heals, stimulates, recharges, re-motivates and aligns the meridians of the body – directing energy to where it is needed most. It promotes truth and forgiveness. Topaz brings joy, generosity, abundance and good health. It is known as a stone of love and good fortune. It releases tension, inducing relaxation. Topaz promotes openness and honesty, self-realization and self-control. It aids problem-solving and assists in expressing ideas. Stabilizes the emotions, making you receptive to love from all sources.

Topaz aids digestion and combats eating disorders, such as anorexia and bulimia. It fortifies the nerves and stimulates the metabolism.

Tourmaline

Tourmaline comes in various colors. A single crystal may contain multiple colors. Tourmaline is a boron silicate gemstone that may contain any of a number of other elements, giving it a chemical formula of $(Ca,K,Na,[])(Al,Fe,Li,Mg,Mn)_3(Al,Cr, Fe,V)_6$

$(BO_3)_3(Si, Al, B)_6O_{18}(OH, F)_4$. It forms trigonal crystals and has a hardness of 7 to 7.5. Tourmaline is often black but may be colorless, red, green, bi-colored, tri-colored, or other colors.

Tourmaline aids in understanding oneself and others. It promotes self-confidence and diminishes fear. Tourmaline attracts inspiration, compassion, tolerance and prosperity. It balances the right-left sides of the brain. Helps treat paranoia, overcomes dyslexia and improves hand-eye coordination. Tourmaline releases tension, making it helpful for spinal adjustments. It balances male-female energy within the body. Enhances energy and removes blockages.

Turquoise
Turquoise is an opaque gem, often seen in shades of blue, green, and yellow.

Like pearl, turquoise is an opaque gemstone. It is a blue to green (sometimes yellow) mineral consisting of hydrated copper and aluminum phosphate. Its chemical formula is $CuAl_6(PO_4)_4(OH)_8 \cdot 4H_2O$. Turquoise belongs to the triclinic crystal system and is a relatively soft gem, with a Mohs hardness of 5 to 6.

Turquoise is a purification stone. It dispels negative energy and can be worn to protect against outside influences or pollutants in the atmosphere. Turquoise balances and aligns all the chakras, stabilizing mood swings and instilling inner calm. It is excellent for depression and exhaustion; it also has the power to prevent panic attacks. Turquoise promotes self-realization and assists creative problem solving. It is a symbol of friendship and stimulates romantic love.

Turquoise aids in the absorption of nutrients, enhances the immune system, stimulates the regeneration of tissue, and heals the whole body. It contains anti-inflammatory and detoxifying effects and alleviates cramps and pain. Turquoise purifies lungs, soothes and clears sore throats, and heals the eyes, including cataracts. It neutralizes over acidity, benefits rheumatism, gout, stomach problems, and viral infections.

Zircon

Zircon comes in a wide range of colors. The element phosphorus is very important for many facets of life. Thus, phosphate minerals in which phosphorus is oxidized in the phosphate group PO4 are part of a tight geochemical cycle that includes the biosphere, rather like the carbon cycle.

Zircon Gemstone. Magical properties -It consists of concentrating energy into sharp focus and gifting its wearer a great feeling of confidence and strength. It helps to erase negativity from inside and the feeling of deception. It helps the person to uncover lies and look for the facts for a better life.

Sulfide Minerals

Bornite

The sulfide minerals represent higher temperatures and a slightly deeper setting than the sulfate minerals, which reflect the oxygen-rich environment near the Earth's surface. Sulfides occur as primary accessory minerals in many different igneous rocks and in deep hydrothermal deposits that are closely related to igneous intrusions. Sulfides also occur in metamorphic rocks where sulfate minerals are broken down by heat and pressure, and in sedimentary rocks where they are formed by the action of sulfate-reducing bacteria. The sulfide mineral specimens you see in rock shops come from the deep levels of mines, and most display a metallic luster.

Bornite (Cu_5FeS_4) is one of the lesser copper ore minerals, but its color makes it highly collectible. (more below)

Bornite stands out for the amazing metallic blue-green color it turns after exposure to the air. That gives Bornite the nickname peacock ore. Bornite has a Mohs hardness of 3 and a dark gray streak.

Copper sulfides are a closely related mineral group, and they often occur together. In this bornite specimen are also bits of golden metallic chalcopyrite ($CuFeS_2$) and areas of dark-gray chalcocite (Cu_2S). The white matrix is calcite. I

Chalcopyrite

Chalcopyrite, $CuFeS_2$, is the most important ore mineral of copper. Chalcopyrite (KAL-co-PIE-rite) usually occurs in massive form, like this specimen, rather than in crystals, but its crystals are unusual among the sulfides in having a shape like a four-sided pyramid (technically they are scalenohedral). It has a Mohs hardness of 3.5 to 4, a metallic luster, a greenish black streak and a golden color that is commonly tarnished in various hues (though not the brilliant blue of bornite). Chalcopyrite is softer and yellower than pyrite, more brittle than gold. It is often mixed with pyrite.

Chalcopyrite may have various amounts of silver in place of the copper, gallium or indium in place of the iron, and selenium in place of the sulfur. Thus, these metals are all byproducts of copper production.

Cinnabar

Cinnabar, mercury sulfide (HgS), is the principal ore of mercury. (more below)

Cinnabar is very dense, 8.1 times as dense as water, has a distinctive red streak and has hardness 2.5, barely scratchable by the fingernail. There are very few minerals that might be confused with cinnabar, but realgar is softer and cuprite is harder.

Cinnabar is deposited near the Earth's surface from hot solutions that have risen from bodies of magma far below. This crystalline crust, about 3 centimeters long, comes from Lake County, California, a volcanic area where mercury was mined until recently.

Galena

Galena is lead sulfide, PBS, and is the most important ore of lead.

Galena is a soft mineral of Mohs hardness of 2.5, a dark-gray streak and a high density, around 7.5 times that of water. Sometimes galena is bluish gray, but mostly it's straight gray.

Galena has a strong cubic cleavage that is apparent even in massive specimens. Its luster is very bright and metallic. Good pieces of this striking mineral are available in any rock shop and in occurrences around the world. This galena specimen is from the Sullivan mine in Kimberley, British Columbia.

Galena forms in low- and medium-temperature ore veins, along with other sulfide minerals, carbonate minerals, and quartz. These can be found in igneous or sedimentary rocks. It often contains silver as an impurity, and silver is an important byproduct of the lead industry.

Marcasite

Marcasite is iron sulfide or FeS_2, the same as pyrite, but with a different crystal structure. Marcasite forms at relatively low temperatures in chalk rocks as well as in hydrothermal veins that also host zinc and lead minerals. It doesn't form the cubes or pyritohedron typical of pyrite, instead forming groups of spearhead-shaped twin crystals also called cockscomb aggregates. When it has a radiating habit, it forms "dollars," crusts and round nodules like this, made of radiating thin crystals. It has a lighter brass color than pyrite on a fresh face, but it tarnishes darker than pyrite, and its streak is gray whereas pyrite may have a greenish-black streak.

Marcasite tends to be unstable, often disintegrating as its decomposition creates sulfuric acid.

Metacinnabar

Metacinnabar is mercury sulfide (HgS), like cinnabar, but it takes a different crystal form and is stable at temperatures above 600°C (or when zinc is present). It is metallic gray and forms blocky crystals.

Molybdenite

Molybdenite (Mo-LIB-denite) is the only mineral that might be confused with graphite. It's dark, it's very soft (Mohs hardness 1 to 1.5) with a greasy feel, and it forms hexagonal crystals like graphite. It even leaves black marks on paper like graphite. But its color is lighter and more metallic, its mica-like cleavage flakes are flexible, and you may see a glimpse of blue or purple between its cleavage flakes.

Molybdenum is necessary for life in trace amounts, because some vital enzymes require an atom of molybdenum to fix nitrogen to build proteins. It's a star player in the new biogeochemical discipline called metallomics.

Pyrite

Pyrite, iron sulfide (FeS_2), is a common mineral in many rocks. Geochemically speaking, pyrite is the most important sulfur-containing mineral. (more below)

Pyrite occurs in this specimen in relatively large grains associated with quartz and milky-blue feldspar. Pyrite has a Mohs hardness of 6, a brass-yellow color and a greenish black streak.

Pyrite resembles gold slightly, but gold is much heavier and much softer, and it never shows the broken faces that you see in these grains. Only a fool would mistake it for gold, which is why pyrite is also known as fool's gold. Still, it's pretty, it's an important geochemical indicator, and in some places, pyrite really does include silver and gold as a contaminant.

Pyrite "dollars" with a radiating habit are often found for sale at rock shows. They are nodules of pyrite crystals that grew between layers of shale or coal.

Pyrite also readily forms crystals, either cubic or the 12-sided forms called pyritohedron. And blocky pyrite crystals are commonly found in slate and phyllite.

Sphalerite

Most often sphalerite is reddish-brown, but it can range from black to (in rare cases) clear. Dark specimens can appear somewhat metallic in luster, but otherwise its luster can be described as resinous or adamantine. Its Mohs hardness is 3.5 to 4. It commonly occurs as tetrahedral crystals or cubes as well as in granular or massive form.

Sphalerite can be found in many ore veins of sulfide minerals, commonly associated with galena and pyrite. Miners call sphalerite "jack," "blackjack," or "zinc blende." Its impurities of gallium, indium and cadmium make sphalerite a major ore of those metals.

Sphalerite has some interesting properties. It has excellent dodecahedral cleavage, which means that with careful hammer work you can chip it into nice 12-sided pieces. Some specimens fluoresce with an orange hue in ultraviolet light; these also display triboluminescence, emitting orange flashes when stroked with a knife.

Apatite

Apatite ($Ca_5(PO_4)_3F$) is a key part of the phosphorus cycle. It is widespread but uncommon in igneous and metamorphic rocks.

Apatite is a family of minerals centered around fluorapatite, or calcium phosphate with a bit of fluorine, with the formula $Ca_5(PO_4)_3F$. Other members of the apatite group have chlorine or hydroxyl that take the place of the fluorine; silicon, arsenic or vanadium replace the phosphorus (and carbonate replace the phosphate group); and strontium, lead, and other elements substitute for the calcium. The general formula for the apatite group is thus $(Ca,Sr,Pb)_5[(P,As,V,Si)O_4]_3(F,Cl,OH)$. Because fluorapatite makes up the framework of teeth and bones, we have a dietary need for fluorine, phosphorus, and calcium.

This element is usually green to blue, but its colors and crystal forms vary. Apatite can be mistaken for beryl, tourmaline, and other minerals (its name comes from the Greek "apate," or deceit). It is most noticeable in pegmatites, where large crystals of even rare minerals are found. The main test of apatite is by its hardness, which is a 5 on the Mohs scale. Apatite can be cut as a gemstone, but it is relatively soft.

Apatite also makes up sedimentary beds of phosphate rock. There it is a white or brownish earthy mass, and the mineral must be detected by chemical tests.

Lazulite

Lazulite, $MgAl_2(PO_4)_2(OH)_2$, is found in pegmatites, high-temperature veins, and metamorphic rocks.

The color of lazulite ranges from azure- to violet-blue and bluish-green. It's the magnesium end member of a series with the iron-bearing scorzalite, which is very dark blue. Crystals are rare and wedge-shaped; gemmy specimens are even rarer. Typically you'll see small bits without good crystal form. Its Mohs hardness rating is 5.5 to 6.

Lazulite can be confused with lazurite, but that mineral is associated with pyrite and occurs in metamorphosed limestones. It is the official gemstone of the Yukon.

Pyromorphite

Pyromorphite is a lead phosphate, $Pb_5(PO_4)_3Cl$, found around the oxidized edges of lead deposits. It is occasionally an ore of lead.

Pyromorphite is part of the apatite group of minerals. It forms hexagonal crystals and ranges in color from white to gray through yellow and brown but is usually green. It is soft (Mohs hardness 3) and very dense, like most lead-bearing minerals.

Variscite

Variscite is a hydrous aluminum phosphate, $Al(H_2O)_2(PO_4)$, with a Mohs hardness of around 4.

It forms as a secondary mineral near the surface in places where clay minerals and phosphate minerals occur together. As these minerals break down, variscite forms in massive veins or crusts. Crystals are small and very rare. Variscite is a popular specimen in rock shops.

This variscite specimen comes from Utah, probably the Lucin locality. You might see it called lucinite or possibly utahlite. It looks like turquoise and is used the same way in jewelry, as cabochons or carved figures. It has what's called a porcelaneous luster, which is somewhere between waxy and vitreous.

Variscite has a sister mineral called strengite, which has iron where variscite has aluminum. You might expect there to be intermediate mixtures, but only one such locality is known, in Brazil. Usually strengite occurs in iron mines or in pegmatites, which are very different settings from the altered phosphate beds where variscite is found.

ROCK DUST FOR GROWING

Azomite is some of the best rock dust on the market. Mixed with worm castings and compost, azomite feeds plants what they need for robust growth and fruit production. Great for all gardens but especially for those with depleted soils. Put back all those trace minerals into the soil!

GPO Can It Re-mineralizers the Earth?

Rock dust is a very popular soil additive especially with organic and permaculture groups. It is full of nutrients and adding it to soil will replenish all the nutrients that agriculture has taken out of our soil. This process of adding nutrients back to soil is known as mineralization.

This seems to make a lot of sense. We remove food from the land, and the food contains lots of minerals. At some point we need to put them back into the soil or else we will have soil that won't grow anything.

Azomite – a common brand of rock dust

What is Rock Dust?

The simple definition is that rock dust, also known as rock powder and rock flour, is pulverized rock. It can be man-made or occur naturally. Cutting granite for commercial use produces granite dust. Glaciers naturally produce glacial rock dust. Rock dust is also found near ancient volcanoes and consists of basalt rock. To be effective the rock needs to be ground into a very fine powder. That way it is more easily used by microorganisms and decomposed by environmental elements. Two common forms of rock, namely limestone and phosphate rock have been used for a long time to amend soil. Although these products are correctly called rock dust, they are usually not included when gardeners talk about rock dust, and I will exclude them from this post. Some commercial products call themselves a fertilizer and I even found one that was labeled like a fertilizer showing an NPK of 0-0-1, but by most legal definitions rock dust does not contain enough NPK to qualify as a fertilizer.

Helps restore the correct mineral balance in soil

To be true, this would mean that soil has some kind of "correct balance" to begin with and that this balance is important for plant growth. It turns out that there are many kinds of soil, and they vary widely in their mineral composition. There are plants that are adapted to and grow on just about any soil. There is no such thing as a "correct mineral balance".

Analysis reports show Lanthanum (La), Cerium (Ce) and Praseodymium (PR) at 644 ppm

These are *rare earth* elements, which makes it sound as if you would want them in your soil – who does not want *rare* stuff? I have heard of the first two, but not praseodymium – I must have been away the day we did experiments with it! The claims go on to say, "These elements act as cofactors for the methanol dehydrogenase of the bacterium *Methylacidiphilum fumariolicum.*" So what is this important bacterium? *Methylacidiphilum fumariolicum* is an autotrophic bacteria, first described in 2007 growing on volcanic pools near Naples, Italy. It grows in mud at temperatures between 50 °C – 60°C (about 130 °F) and an acidic pH of 2–5.That may be true, but there seems to be no published research to show that paramagnetic rock has any effect on plant growth. However, many pseudoscience groups do make such claims. Rock dust does contain a lot of minerals. I have seen claims ranging from 60 up to 90 different minerals.

How Much Should You Use?

I find that this question can tell you a lot about a product. If rock dust is good for gardens, how much should you use? What happens if you use too much? One site had this recommendation.

3 tons/acre = 14 lb./100 sq. ft. = 1.25 lb./sq. yd.

or

7.5 tons/ha = 750 kg/1000 sq. = 75 kg/100 sq. = 750 grams/1 sq.

But a rate even 8x higher can be used, although it would have to be incorporated into the soil.

You can add anywhere from 3 tons/acre to 24 tons/acre. If 3 was the right number, would 24 not be way too much? Would 24 not burn plants due to the high nutrient load? Only if the product added nutrients to soil

WATER FILTERS

In this section there are some examples of a water filtration system using rocks and gemstones to clean water. Growing Pharms / GPO has our own variation of stones we use for different plants and to induce different effects in the effects people get by partaking in the biomass from the plant. We have our own filters and pumps. This picture are close to some of the private models we use.

Biofilter Waterfall
Cleans water from the skimmer via UV sterilization

Filtration Zone
Planted filter

Retaining Wall

Planted Filter Pump & UV Sterilizers
The pump sucks water from the slotted plastic drain tube under the planted filter and/or a skimmer and pushes it through a UV sterilizer to the waterfall, stream or bottom out flow

Swimming Zone

Plant Filter Medium
A mixture of lava rock & river rock Provides a biofilter medium for the moving water & a growth medium for water plants

Bottom Out Flow
Stirs up water from the bottom to prevent untreated water

Ayer's Rock Ceramic Marbles

Negative Ion Ceramic Marbles
- Emit negative ions for healthy cells
- Help to fight against free radicals
- Boost immune system

Alkaline Ceramic Marbles
- Create Alkaline Water
- Increase pH up to 10.5
- Help the body to absorb minerals better

Tourmaline Ceramic Marbles
- Reduce amount of clusters in water molecule from 20 to 5
- Develop synergy between the Negative Ion and Far-Infrared Ceramic Marbles

Far-Infrared Ceramic Marbles
- Emit Far-Infrared Rays
- Activate metabolic functions
- Increase oxygen content in blood

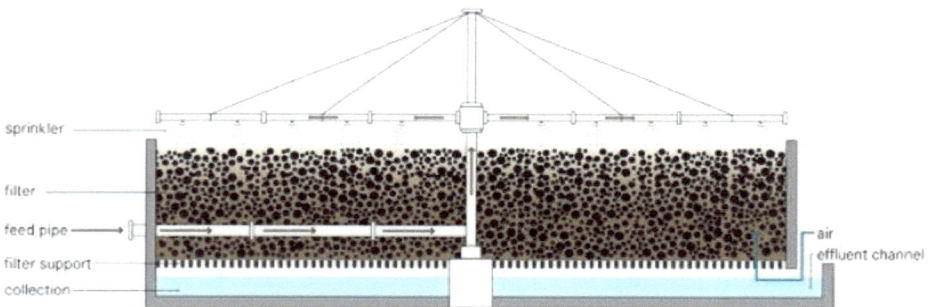

sprinkler

filter

feed pipe

filter support

collection

air

effluent channel

FINISHED PRODUCTS

BIBLIOGRAPHY

Government Resources
NASA is an expert in climate and Earth science.
While its role is not to set climate policy or prescribe
responses or solutions to Bibliography

Government Resources
NASA is an expert in climate and Earth science. While its role is not to set climate policy or prescribe responses or solutions to climate change, its purview does include providing the robust scientific data needed to understand climate change and evaluating the impact of efforts to combat it. NASA then makes this information available to the global community – the public, policy- and decision-makers and scientific and planning agencies around the world. (For more information, see NASA's role.)

The following selected resources from U.S. government organizations provide information about options for responding to climate change.

Data and Information

- DATA.GOV

Climate Data Initiative

Data related to climate change that can help inform and prepare America's communities, businesses and citizens.

- U.S. Climate Resilience Toolkit

U.S. Climate Resilience Toolkit

Provides scientific tools, information and expertise to help people manage their climate-related risks and opportunities and improve their resilience to extreme events.

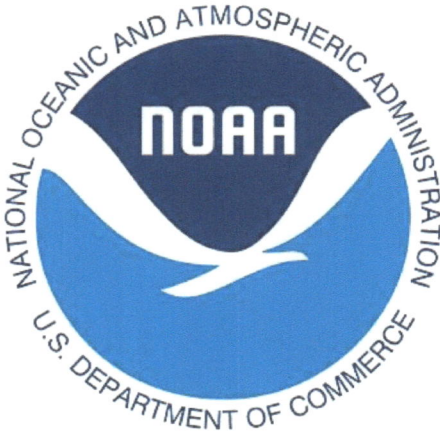

- **National Oceanic and Atmospheric Administration**

From supercomputers and state-of-the-art models to observations and outlooks, this site provides data, tools and information to help people understand and prepare for climate variability and change.

- **National Climate Assessment 2014**

Produced by a team of more than 300 experts and guided by a 60-member Federal Advisory Committee, this report summarizes current and future impacts of climate change on the United States.

- **U.S. Department of Energy**

Describes strategies currently being pursued or considered to reduce carbon emissions and address global climate change.

-
Environmental Protection Agency

Tools for learning and understanding environmental issues and recommendations for greener living.

-
State of California's Climate Change Portal

This site provides many links to state reports on climate change mitigation and adaptation options.

-
U.N. Framework on Climate Change

United Nations Framework Convention on Climate Change On-line newsletter concerning issues about the U.N.'s convention on long-term climate change.

Reducing tillage
With better weed-control solutions, farmers reduce the need to till, decreasing tractor passes over the field and allowing for less soil disruption. This not only curbs greenhouse gas emissions and fossil fuel use, but when soil is left untilled it is better able to store carbon, as well as nutrients and water.

Increasing efficiency

Digital tools and precision agriculture techniques enable farmers to have a more intimate and informed understanding of what's happening in their fields. For example, Bayer is working on software platforms that offer monitoring tools which help farmers use pesticides more efficiently, reducing greenhouse gas emissions and pesticide runoff into water. Precipitation alerts let farmers know which fields may be too wet or too windy to apply pesticides so they can avoid costly treatments and unnecessary fossil fuel use. Satellites and drones provide real-time field health images that enable farmers to identify areas of crop stress or pest infestations, so that corrective actions can be taken quickly and efficiently. These tools help farmers optimize land use to grow enough food on less acreage, offering the potential to reduce the number of acres needed to feed a growing population.

http://www.homehydrosystems.com/mediums/mediums_page.html

by Anne Marie Helmenstine, Ph.D.

https://www.charmsoflight.com

<u>INDEX</u>

www.ingramcontent.com/pod-product-compliance
Lightning Source LLC
Chambersburg PA
CBHW041713200326
41519CB00001B/147